João Granjo

Industrial Energy Management

João Granjo

Industrial Energy Management

LAP LAMBERT Academic Publishing

Impressum / Imprint
Bibliografische Information der Deutschen Nationalbibliothek: Die Deutsche Nationalbibliothek verzeichnet diese Publikation in der Deutschen Nationalbibliografie; detaillierte bibliografische Daten sind im Internet über http://dnb.d-nb.de abrufbar.
Alle in diesem Buch genannten Marken und Produktnamen unterliegen warenzeichen-, marken- oder patentrechtlichem Schutz bzw. sind Warenzeichen oder eingetragene Warenzeichen der jeweiligen Inhaber. Die Wiedergabe von Marken, Produktnamen, Gebrauchsnamen, Handelsnamen, Warenbezeichnungen u.s.w. in diesem Werk berechtigt auch ohne besondere Kennzeichnung nicht zu der Annahme, dass solche Namen im Sinne der Warenzeichen- und Markenschutzgesetzgebung als frei zu betrachten wären und daher von jedermann benutzt werden dürften.

Bibliographic information published by the Deutsche Nationalbibliothek: The Deutsche Nationalbibliothek lists this publication in the Deutsche Nationalbibliografie; detailed bibliographic data are available in the Internet at http://dnb.d-nb.de.
Any brand names and product names mentioned in this book are subject to trademark, brand or patent protection and are trademarks or registered trademarks of their respective holders. The use of brand names, product names, common names, trade names, product descriptions etc. even without a particular marking in this works is in no way to be construed to mean that such names may be regarded as unrestricted in respect of trademark and brand protection legislation and could thus be used by anyone.

Coverbild / Cover image: www.ingimage.com

Verlag / Publisher:
LAP LAMBERT Academic Publishing
ist ein Imprint der / is a trademark of
OmniScriptum GmbH & Co. KG
Heinrich-Böcking-Str. 6-8, 66121 Saarbrücken, Deutschland / Germany
Email: info@lap-publishing.com

Herstellung: siehe letzte Seite /
Printed at: see last page
ISBN: 978-3-659-31388-2

Zugl. / Approved by: Porto, University of Porto, 2012

Acknowledgements

Apesar da língua inglesa dar voz ao trabalho que se apresenta, não poderia deixar de agradecer os vários, e valiosos, contributos que me foram dados na minha língua materna.

Ao Eng. Francisco Leitão, responsável pela gestão de energia da SONAE Indústria, não há palavras para agradecer o seu tempo a mostrar-me o complexo industrial, a partilhar os meus problemas e a mostrar-me a cidade onde, antigamente, não conhecia ninguém e onde agora tenho bons amigos. Um sincero agradecimento.

Ao Prof. Dr. José Nuno Moura Marques Fidalgo, um sentido agradecimento pela orientação, flexibilidade e disponibilidade com que orientou este projeto, é-me impossível imaginar um melhor orientador para este trabalho.

Por último, mas de especial importância, quero agradecer aos meus pais e à minha irmã, que sempre me apoiaram nesta jornada na que culmina com a publicação deste trabalho. Igualmente, e como se de família se tratassem, agradeço aos meus amigos e à minha namorada, pelas horas de convívio que muito ajudaram nas alturas de maior stress e frustração.

A todos os mencionados, e a outros cujos agradecimentos serão dados pessoalmente, obrigado por tudo.

Table of Contents

List of Figures

List of Tables

Abbreviations, Acronyms and Symbols

List of Abbreviations and Acronyms

AC	Alternate Current
BSC	Balanced Scoreboard
e.g.	exempli gratia (meaning "for example")
ESCO	Energy Service Company
FEUP	Faculty of Engineering of the University of Porto
IRR	Internal Rate of Return
kW	Kilowatt
kWh	Kilowatt-hour
kWp	Kilowatt peak
MDF	Medium-Density Fiberboard
MIRR	Modified Internal Rate of Return
MWh	Megawatt-hour
NPV	Net Present Value
PCDM	Produção e Comercialização Derivados de Madeira (Production and commercializing of wood derived products)
PVC	Polyvinyl chloride
toe	Tonne of oil equivalent
VSD	Variable Speed Drive

List of symbols

€	Euros

Chapter 1

Introduction

Energy is becoming increasingly scarce and, as such, it is urgent that all governments, industries and citizens adopt measures to utilize it efficiently. International committees have increased the pressure on governments to enforce the application of electrical efficiency measures: as an example, the European Commission adopted a plan which aims at the reduction of 20% of the consumption of energy in Europe, until the year 2020 [1]! It is even thought that one of biggest world' challenge is to provide energy which is easily accessed and safe, without impacting the environment [2].

Aware of the importance of this matter, SONAE Indústria, one of the largest wood-based panels' producers of the world [3], decided to develop a plan to reduce the specific consumption of their operations. In order to do so, the first step taken at a corporate level was to assign an energy manager in each of its 26 factories. This manager is in charge of developing projects to reduce the consumption of his factory and sharing their projects with the remaining group through monthly video-conference meetings.

That strategic change, which has been imposed since 2009, produced some interesting results: new technologies were introduced, energy was no longer seen as insignificant, efficiency had been improved, costs had been reduced and emissions had decreased. However, SONAE wanted to keep pursuing efficiency and cost reduction, and now it had at their disposable energy managers in each of its factories. Nonetheless, the simple, efficiency improvement projects have all been implemented; new ideas started appearing at a slower pace than before, and the measures that could be taken now are, apparently, the least profitable and, with the present economic crisis, senior management is not easily convinced in making funds available for projects that will that take over 5 years to provide any gains.

To overcome this apparent stall in efficiency improvements SONAE launched two pilot projects in two plants, which, if successful, are expected to be replicated in the remaining factories, spread over 13 countries.

One of these projects is the one this book describes: the performance of the most thorough energy audit ever achieved in one of SONAEs' factories. Energy audits have been routinely performed before, as it is mandatory by the Portuguese legislation for facilities that consume over 1000 toe per year to have an energy audit every six years [4]. The difference is that, this time, the objective is to go into the smallest details, having been assembled a very large team which comprises knowledge from SONAE Indústria, which is the second world's biggest producer of wood board, EDP, which ranks among Europe's major electricity operators, as well as being one of the largest business groups of Portugal, and the University of Porto, the largest Portuguese university by number of enrolled students and one of the most noted research outputs in Portugal [5]. This project is seen as an important improvement driver, which is noticeable not only by the involvement of such a large team, but also due to the funds made available which are about 5 times higher than what it is usually paid for an energy audit.

This way, the following book describes the most relevant issues of this project, and aims at providing the reader with enough information to be able to replicate the achievements and gain an understanding of the complexity an industrial energy audit might impose.

1.1 - Objectives

The project, in which this book is based, took place in the industrial plant established in Mangualde and had the purpose of performing an energy audit, in order to understand the energy needs and, subsequently, manage them and reduce the consumptions. As such, this dissertation has the objective of depicting the project, in order to share the gained *know-how* with the remaining industrial facilities - specially the one's operated by SONAE - and to predict the impact of the energy efficiency measures implementation. As such, it includes the following objectives:

- Theoretical defining an energy audit and defining the different steps it encompasses;
- Characterize the audit process, providing guidelines for project replication, focusing on the most important success drivers :
  - o Audit Planning;
  - o Information Requirements;
  - o Fieldwork;
- Identify situations of energy waste;
- Propose corrective measures for energy waste situations and new technologies to increase efficiency;
- Estimate the outcome of the implementation of the energy efficiency measures;
- Comment future work to further increase efficiency.

To sum up, this book intends to provide the reader with knowledge concerning an industrial energy audit, commenting on the different aspects that impact the outcome of the project, and analyzing the results. It is based on the actual work that was accomplished in SONAE and, as such, it provides both the theoretical perspective that was obtained through the analyzes of the works of the different authors referenced throughout the book, and the practical perspective, gained through the time spent in the factory, witnessing the difficulties and accomplishments that this project had.

1.2 - Document Structure

This report includes seven chapters besides this introductory chapter, where the topic that will be covered throughout this book is portrayed. Concerning the remaining chapters:

Chapter 2 provides a background of the factory in which the energy audit will be performed, defining both the plant and the productive process. It is important as the reader needs to be aware of the activity developed in the plant, in order to understand the challenges and particularities of this project.

Chapter 3 provides a theoretical context concerning the energy audit, explaining the best practices to perform the assessment, and clarifying the different processes that constitute this task. This chapter is relevant as it allows to compare the theoretical best practices with the actual work performed, allowing the reader to develop his own opinion concerning how such a project should be develop.

From chapter 4 to chapter 6, the actual activities for the energy audit are presented. Chapter 4 describes the information requirements for such an exhaustive audit, along with comments concerning the relevance of the information and the issues found while it was being gathered.

Chapter 5 comments on the fieldwork performed in the factory, more precisely it details the procedure followed to determine the most relevant monitoring points, and intends to provide criterion for the replication of the audit in other facilities.

Chapter 6 documents the energy efficiency measures to be implemented that are feasible both from the technical and the financial point of view. Simultaneously, it also refers measures that were analyzed but not implemented due to various reasons. Once again, it aims at providing enough information for the reproduction in other industrial facilities.

Chapter 7 predicts the impact of the implementation of the measures presented in the previous chapter on the future consumptions.

Chapter 8 summarizes the whole book, mentioning the main highlights and difficulties found during this work and discussing other chores to further increase energy efficiency.

Chapter 2

Factory Background

The following chapter offers some information concerning the industrial unit where the energy audit, in which this book is based, was performed. Ranging from a general factory overview to a specific description of the productive process, this chapter intends to provide the reader with background information to allow a better understanding of the audit process, and energy efficiency measures, described subsequently.

2.1 - General Factory Information

The plant in which the audit was performed is a plant that produces mainly Medium Density Fiberboard and wood veneer surfacing. It employs around 184 permanent employees and occupies an area of 27 ha, which, to put under perspective, is the area of about 32 football fields. The general factory information is summarized on the following table:

Table 2.1 General Factory Information

Firm	SONAE Indústria P.C.D.M., S.A.	
Main Activity	Production of Medium Density Fiberboard and wood veneer surfacing	
Area	27 ha	
Nr. of Employees	184	
Operating Hours	MDF Production	Continued operation, 24h/day
	Remaining activities	3 shifts a day, 5 days a week.

2.2 - Location

The plant is located in Água Levada in the county of Mangualde, district of Viseu, as shown inFigure 2.1 Location of SONAE Indústria Mangualde PCDM, SA Figure 2.1 Figure 2.2:

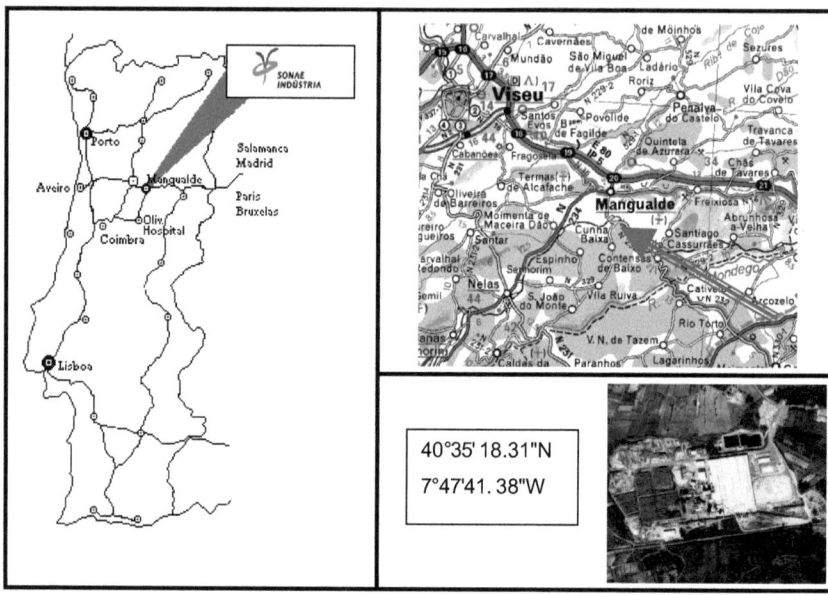

40°35'18.31"N
7°47'41.38"W

Figure 2.1 Location of SONAE Indústria Mangualde PCDM, SA

2.3 - Historic Overview

After the installation of the factory in 1987, the factory suffered several improvements. The changes and developments it undertook up to this moment are presented in the following bullets [6]:

1987   SONAE Indústria launches the project of an MDF plant at Mangualde, under the name *SIAF - Sociedade de Iniciativa e Aproveitamentos Florestais, SA* (a forestry company created in 1946 by the Swedish Match Group that was later involved in wood-based panels and doors manufacturing and that was bought by SONAE Indústria at this time).

1988   Building construction start up, and startup of mechanical assembly.

1989   Company head-office moves to Mangualde and production of the first Sonaepan board, trade mark of MDF produced by SIAF.

1990 On the 8th of October there's the official ceremony marking the opening of the Mangualde Industrial unit.

1992 SONAE Indústria – SGPS, SA, gets the control of 100% of SIAF bonds.

1993 The plant obtains the ISO 90000 certification and later that year, with a new diesel power generator unit, the unit becomes energy self-sufficient.

1994 Approval of a new MDF production line and start-up of that line installation (line II).

1995 Official opening of the second MDF production line and production of the first MDF board in line II. Later that year, the unit was recognized has a company responsible for electrical power and heat generation and supply.

1996 Start-up of the 2nd stage of biological treatment of the waste water treatment unit.

1997 Start-up of SIAF-Energy, SA, a company that manages the selling of the electrical production generated in the plant.

1998 Start-up of wood veneer surfacing line I.

2000 Start-up of wood veneer surfacing line II and, following the reorganization process of SONAE Indústria, the plant becomes part of *Casca-Sociedade de Revestimentos, SA*.

2004 The MDF plant becomes part of SONAE Indústria – Produção e Comercialização de Derivados de Madeira, SA.

2005 The former ESSO Portuguesa Lda paraffin emulsions plant, located inside the Mangualde site, is integrated in the SONAE Indústria operation.

2007 Certification of the Environmental Management System according to standard NP EN ISO 14001:2004 in 20-06-2007 (Certification n. PT-2007/AMB.0313 issued by APCER – Associação Portuguesa de Certificação).

2008 Start-up of the third stage of the waste water treatment process, enabling the reuse of the treated water.

Figure 2.2 Aerial picture of the factory on the year 2012

2.4 - Productive Process

As stated before, there are mainly two products being produced in the factory: the MDF board and the wood veneer surfacing. Both production processes have a certain complexity which could lead to a much extended description. But, since the purpose of the chapter is to provide the reader with a simple production overview, only the general process will be analyzed , and commented upon.

2.4.1 - Flowchart of the Productive Process of MDF

The following page presents a flowchart of the productive process of the MDF board. It is a simple overview, but offers a clear idea of process. On the top of the flowchart, the production of electricity and thermal energy, through the cogeneration plant, is displayed. Bellow it is presented the actual process to create a MDF board, starting from the wood chipper. It is interesting to notice how connected the cogeneration plant is with the production process, meaning that the behavior of the cogeneration plant (the thermal energy it provides) directly affects the production of the MDF. The same analysis will not be presented for wood veneer surfacing since it is a simpler process and a description is sufficient.

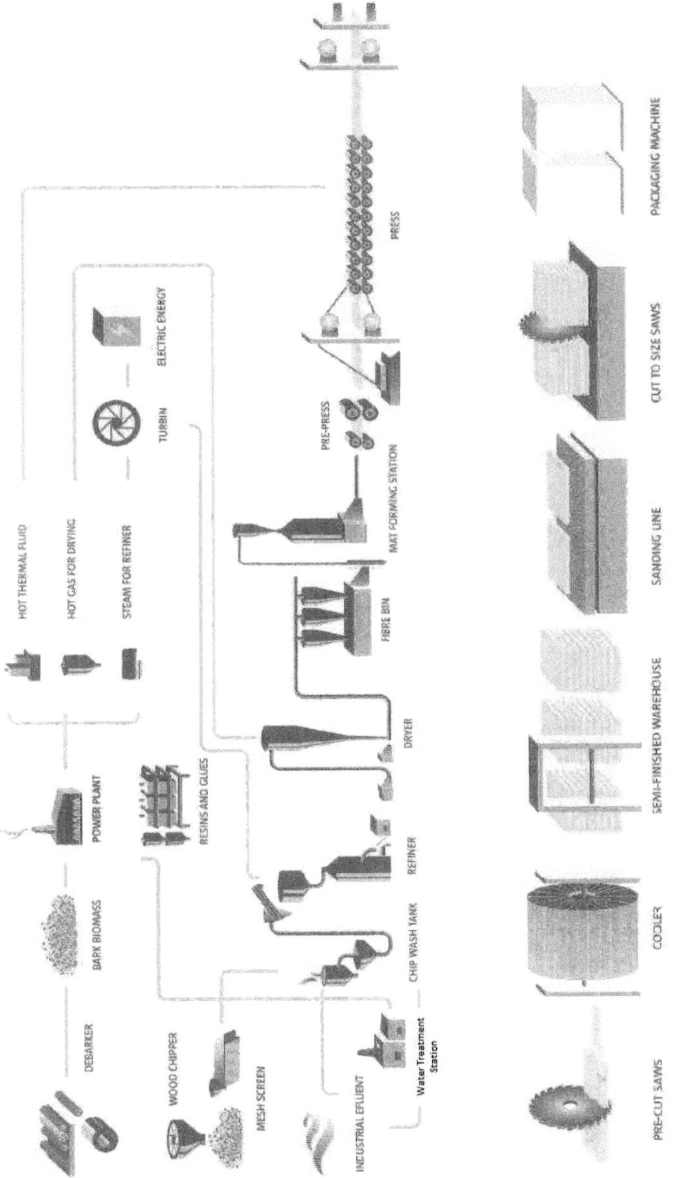

17

Figure 2.3 Flowchart of the MDF productive process - part 1 [6]

2.4.2 - Description of the MDF Productive Process

Concerning the production of the MDF board, named Sonaepan board, it is an industrial process that transforms wood in various shapes (logs, slabs and chips) into compact and dimensionally stable board. These can be cut, profiled, sanded, painted and surfaced with wood veneer, PVC, paper or any other type of finishing, just like wood [3]. The main production stages are the following:

1. Chip Production: Chips (small wood strands) are produced in a blade wood chipper, that processes debarked logs and slabs, mainly from pine wood.
2. Chip Screening: Mechanical screening and separation of smaller and larger chips. The smaller chips are used as biomass for the cogeneration plant and the larger chips are redirected to the wood chipper again, to be made smaller.
3. Chip Washing: After the screening, the chip is washed with hot water to remove sand and dirt stuck on the wood pieces. The water used in this process is, afterwards, cleaned on the water treatment plant.
4. Chip Cooking: wood chips are "cooked" with steam in a pressurized digester, to enable an easier refining operation.
5. Refining: The wood chips, which are now softer due to the "cooking process", are mechanically separated into wood fibers, by passing the chips between the metal discs of the refiner. This step is the most energy consuming step as each refiner has a 4MW motor and is working continuously.
6. Glue Spreading: Controlled dosage of the glue is mixed into the fiber. The glue is a mixture of resin and other chemicals in optimized proportions, which afterwards will act as the fiber bonding agent.
7. Drying: The exceeding water content of the fiber is thermally extracted with hot air, being released as water steam through the dryer outlet.
8. Mat forming: Fibers are uniformly deposed as a mat on a conveyor belt.
9. Pressing: The fiber mat is compressed in a *multidaylight* press (on the production line I), or in a continuous press (on the production line II), by means of pressure and temperature. When the glue curing temperature is reached, its polymerisation will bond the fibers and give consistency to the board.
10. Sanding: The board surface is finished through the passage of several sanding machines. The sanding grit gets thinner and thinner from one machine to another, and removes the less dense external overlay of the board.
11. Cutting: The board is cut to achieve the market required standard sizes, according to the customer requirements.
12. Packaging: Board packing and pallet assembly, with all the components that assure the pallet protection and identification, until its arrival to the customer.

2.4.3 - Description of the Wood Veneer Surfacing Productive Process

The wood veneer surfacing, called Lamipan, production is a simpler process. It is also a much smaller energy consumer. To obtain a pack of wood veneer the boards go through the following process:

1. Line inlet: The first operation is to feed the lines with the raw materials: particleboard or MDF and the wood veneer. The veneer had previously been prepared in the lay-ons plant, where cutting and edge gluing of wood veneer is made, to get the lay-on dimensions required.
2. Gluing: Uniform distribution of the glue over the board in both the upper and lower surfaces.
3. Wood veneer lay down: In this step the positioning of wood veneer is done. Both up and down veneer lay-ons are positioned by a mechanical system, which uses vacuum heads to move the lay-ons.
4. Pressing: The board and lay-ons are now heat-pressed together until the glue reaches a reactive temperature and, consequently, glues the wood panels to the core board.
5. Edge trimming: The wood veneer that is oversized in length or width is cut off.
6. Sanding: The boards' surface is sanded to provide it with a smoother exterior.
7. Grading: The boards are classified according to their quality.
8. Packaging: Board packing and pallet assembly, with all the components that assure the pallet protection and identification, until its arrival to the customer.

In the present situation, due to the lack of customer demand, the wood veneer surfacing is only partly functional, but is expected that in the following years, if the market responds as predicted, the production line starts operating at full capacity.

2.5 - Description of the Energy Consumption

Splitting the energy consumption into the different sources, and taking as reference the year of 2010, one can notice that electrical energy is the main consumer, as shown in Figure 2.4, accounting for around 80% of all the energy consumed.

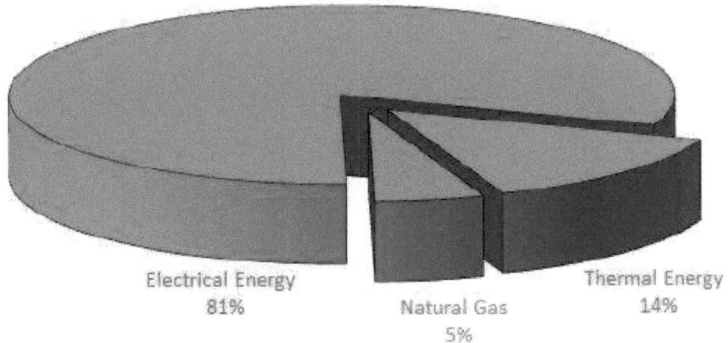

Figure 2.4 Energy consumption split into the different sources [13]

It is possible to conclude, at this point, that the audit should focus primarily on electrical consumption as it is the most relevant driver for costs, and the most widespread energy on the factory. Nonetheless, taking into account that in the previous years this source of energy was the most managed, several improvements were already accomplished, which might mean that there are several opportunities present in the other energy sources.

Concerning the electrical energy, the most significant consumers are the motor driven systems, in particular the refiner, the compressed air systems and the air-driven systems. Regarding the natural gas, it is used as a back-up thermal energy for the productive process, used when thermal energy from the cogeneration plant is not available or when this thermal energy is too "dirty" for the making of premium products. Lastly, the thermal energy is used for the productive process, especially for drying the wood-fiber and for heating the presses.

2.5.1 - Past Works on Energy Efficiency

As stated before, several energy efficiency improvement measures have been undertaken in this factory. Mainly, the measures focused on three areas: Idle reduction time, insertion of VSD in variable load motors and the insertion of a daily process which is called "Energy monitoring, targeting and reporting" and that aims at increasing the awareness for correct energy usage, through daily monitoring on the most significant equipment and daily discussions when abnormal energy consumptions are detected.

Through these three areas, a process for continuous improvement was created which is still undergoing. This way, it was able to obtain a decrease of 5.1% in the specific consumption when comparing the consumption from 2009 to 2011, as is highlighted in Figure 2.5.

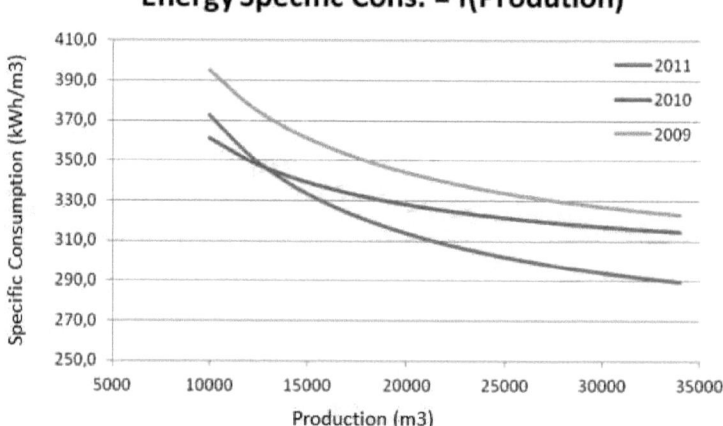

Figure 2.5 Specific Energy Consumption as a function of the production.

Chapter 3

Theoretical Basis of Energy Audits

This chapter provides a characterization of the audit process and advices practices, from several authors, for a successful project. It is an important chapter as it will allow the reader to compare what is advised from a theoretical point of view to what was truly done in the project, and then conclude concerning the value of the practices presented.

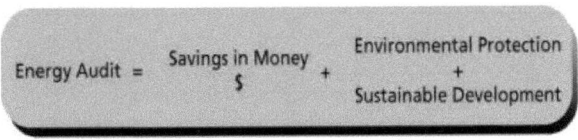

Figure 3.1 Energy audit split into its objectives [8]

3.1 - Characterization of Energy Audits and Efficiency Measures

An energy audit is defined as a detailed examination of the energy usage of a given installation [7]. Through the audit process, it is possible to understand when and where the energy is used, what the efficiencies of the different processes are and where energy waste is happening. In other words, it is an examination of an energy consuming system to ensure that the energy is being used efficiently. Furthermore, it provides solutions for situations of energy waste and attempts to pinpoint those situations.

There are two types of energy audits defined, which are differentiated by the complexity of the work involved in performing them: one is denominated simple audit and, the other one, is named complex audit.

Concerning the simple audit, it comprises the visual observation and a superficial data gathering in order to diagnose the energetic situation of the installation. It is more common to find this type of audits in residential facilities. Regarding the complex audit, it comprises a thorough analysis of the energy usage, detailing the energy utilized in each sector and process. It is the common audit performed on industrial facilities and it is the kind of audit that this book depicts, being that, from now on, the term audit effectively means a complex energy audit. Both of them are useful in providing measures to reduce the energy consumption, and in aiding the operators of the facility to use energy rationally, being differed only due to the scope of each: the simple audit is much more superficial.

Regarding the energy efficiency measures advised in any audit, they are divided in 3 categories in regard to their pay-off time [8] and to their impact on the productive process, as shown in Table 3.1 Categories for energy efficiency measures.

Table 3.1 Categories for energy efficiency measures

Category of Energy efficiency Measure	Capital Cost
Category 1 (named *quick-wins* in SONAE)	Pay-off occurs earlier than 6 months after the implementation. Usually involving idle reduction measures such as turning off non useful machinery.
Category 2	Involves low cost investment with some minor disruption of the operations. Pay-off occurs later than 6 months and earlier than 3 years.
Category 3	Involves relatively high capital cost investment or much disruption of the operations. Pay-off occurs after 3 or more years.

3.2 - Methodology

Different authors refer different methodologies concerning how the audit should be conducted. Nonetheless, all agree that it involves the process of undertaking chores in a well-defined and sequential manner, meaning that the planning stage is crucial [9]. Concerning the chores, it is agreed that they include:

- a detailed analysis of the electrical energy payments of the preceding year;
- a thorough analysis of the equipment present in the facility that produces or consumes electrical and thermal energy, exploring both their operation conditions and the maintenance activities performed on them;
- identification of non-useful energy usage;
- a list of measures in order to improve energy efficiency, that are feasible according to both technical and financial criteria.

Regarding the actual process to achieve the previously mentioned chores, it obviously depends on the size and complexity of the audit, and of the purpose it encompasses. Nonetheless, it includes at least four different phases: planning, fieldwork, data analysis, and elaboration of the report with conclusions and recommendations.

3.2.1 - Planning

The planning stage of any audit constitutes one critical stage, has it directly impacts the efficiency of the remaining processes. Amongst the various chores to be undertaken in this stage, some are important to highlight due to their importance, specifically:
- the determination of the scope of the project;
- the clear statement of the objectives of the audit;
- the schedule, with tollgates and milestones;
- the audit team responsibilities;
- the information requirements.

Without a project plan, due to the high number of activities that should be undertaken, it is more likely that some activities will be forgotten, and that the audit will not be as trustworthy as it could be.

3.2.2 - Field work

The fieldwork includes two main chores:
- Defining the energy consuming equipment that should be monitored
- Installing the meters;

Naturally, the activities included in each of these tasks differ greatly from one industry to another, and therefore nothing can be concluded concerning the time each task might take.

Concerning the equipment required, it will be detailed on 3.3, nonetheless auditing best practices advice the use of portable measurement equipment. This is due to the fact that most industries are not equipped with proper measurement devices with acceptable margin errors.

The outcome of the audit depends greatly on the quality of the work developed on this stage. It is advised that, besides education in energy systems, the auditor possesses experience in auditing before attempting an industrial facility audit due to the increased complexity such facility displays. Above all, it is required that the auditor maintains full awareness in the development of the fieldwork, in order to identify all the possible opportunities for improvement.

3.2.3 - Data Analysis

After the field work, the auditors shall organize and process all the information collected in the preceding stages. This analysis shall produce a number of metrics and indicators, of a quantitative nature, which will allow the assessment of electrical efficiency performance [6].

The most advisable practice is to segments the energy consumption per type of equipment, per type of product, and per factory sector. Concerning the most energy consuming machines, they should be compared to the most energy efficient options in the market in order to understand the savings that could be obtained.

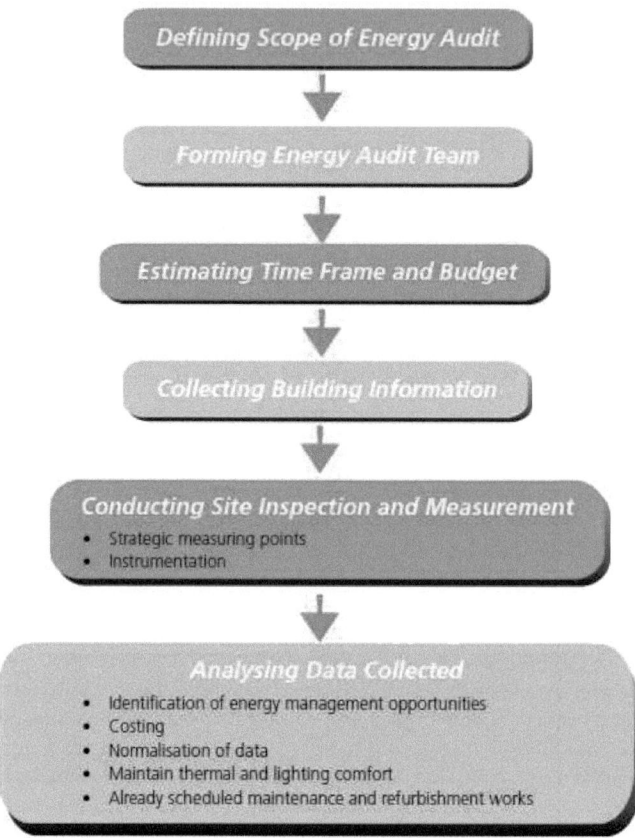

Figure 3.2 Flowchart concerning the activities of an energy audit [8]

3.3 - Equipment Requirements

As stated before, monitoring equipment must be implemented onto the strategic points to measure the energy consumptions. Taking into account the several states in which energy can be found, different metering equipment is required. The next table provides an overview of the equipment for an energy audit.

Table 3.2 Equipment for a detailed energy audit

Instruments	Measured parameters	
Voltmeter	Voltage	Electrical
Ammeter	Current	
Ohmmeter	Resistance	
Multi-meter	Voltage, Current, Resistance	
Wattmeter	Active Power	
Power factor meter	Power Factor	
Light meter (lux meter)	Lighting level in lux (illuminance / illumination level)	
Power quality analyzer	Harmonic contents & Other electrical parameters	
Thermo-graphic scanner/camera	Conductor temperature in °C & Temperature images of overheating conductors (particularly at connection points)	
Thermometer	Dry bulb temperature in °C	Temperature
Sling psychrometer (thermometer)	Both dry and wet bulb temperature in °C	
Portable electronic thermometer		
Infrared remote temperature sensing gun	Useful to sense energy losses due to improper insulation or leakage	
Digital thermometer with temperature probe	Temperature inside a stream of normally hot air/steam (platinum probe for temperature from 0 to 100°C, and thermocouple probe for temperatures as much as 1200°C)	
Hair hygrometer	Humidity/wet bulb temperature	Humidity
Digital thermometer	Humidity/wet bulb temperature	
Pitotstatic tube manometer	Air flow pressure and velocity	Pressure and Velocity
Digital type anemometer with probe	Air flow velocity and pressure	
Vane type anemometer	Air velocity through a coil, air intake, or discharge, for flows that are not dynamically unstable	
Hood type anemometer	Flow rate of air grille	
Pressure gauge	Liquid pressure	
Ultrasonic flow meter with pipe clamps	Liquid flow/velocity	

Instruments	Measured parameters	
Exhaust gas analyzer with probe	Boiler exhaust temperature, O2, CO, CO2 and NOx contents	Miscellaneous
Refrigerant gas leakage tester	Detect refrigerant leakage	
Ultrasonic leak detector	Detect compressed air leakage	
Steam leak detector	Steam leakage, usually for steam trap	
Tachometer	Rotation speed	

Chapter 4

Information Requirements

From this chapter onwards the actual work done in SONAE will be presented. Similarly to what is advised in chapter 3, this process started with several information requirements.

Preceding the actual field-work, it is important to gather as much information as possible regarding the energy consumption in the factory, in order to pin-point the most relevant causes for the energy use. In a smaller facility this task would not be very complicated, but in such a complex unit, where constant improvements are done due to productive needs, it can be a very time-consuming and difficult procedure, especially because the changes performed are not always properly documented.

As such, the following chapter presents the information that was required for this project, in order to provide an information checklist for a new audit project to be performed in another industrial facility. For this particular project, all the data was acquired through archive research, factory surveying and contacting equipment suppliers.

4.1 - General Factory Information

For organizational and planning purposes, general information concerning the site in which the audit will be performed must be obtained. Table 4.1 summarizes this information:

Table 4.1 General Factory Information

Item	Comment
Identification of the company	The best practice would be for the audit team to provide a template where all the relevant information could be inserted.
Topographic plant of the factory	Ideally, the plant would identify the sectors and main equipment.

4.2 - Energy Consumption

Historical data concerning the energy consumption of the facility allows for the auditors to assess the current energy consumption and the seasonable variation of consumption due to different productive behaviors (or maintenance periods). As an example, in this facility, SONAE Indústria Mangualde, the maintenance is usually done in August, and, as such, the energy consumed in that period does not reflect the normal behavior of the factory. Upon noticing this, through the analysis of the historical data, the schedule for the audit was changed. If this had not been identified, the project would have followed the initial plan and the information elicited would have been useless. Table 4.2 summarizes which items should be provided.

Table 4.2 Energy Consumption Information

Item	Comment
Written consent to access energy consumption information of the energy provider	The best practice would be for the audit team to provide a document that would allow this action, upon being signed.
Receipts of all forms of energy used in the factory: electricity, propane, butane, diesel, etc.	Ideally, the receipts should refer to the last 3 years. If it is not possible, at least the receipts of the last 12 months should be provided.

4.3 - Productive Process

The particularities of the productive process impact both the conclusions of the audit and the fieldwork that should be performed. As such, information concerning the process should be provided. Simultaneously, to correctly interpret the energy receipts provided, it is important to also know the production volume, to relate the energy usage with the production needs. Table 4.3 summarizes the information requirements concerning this subject.

Table 4.3 Productive Process Information

Item	Comment
Description and flow chart of the productive process	A summarized description of the productive process is important for the auditor team to gain an understanding of the particularities that impact the productive process
Monthly production volume	In order to better understand the energy consumption and to understand seasonal needs of the factory, a monthly production volume description is required.

4.4 - Motor Driven Systems

Motor-driven systems are the most significant consumers in most industries; as such, information concerning this equipment is of unique importance. Since a facility such as the one analyzed can easily contain over two thousand motors, it is important that it possesses a centralized database listing all the equipment, allowing the access to motors' data in a quick and precise way. Table 4.4 summarizes the information to be gathered concerning this matter and comments on how to do it.

Table 4.4 Motor Driven Systems Information

Item	Comment
List of all the motors in the factory	Often this type of information is difficult to obtain due to the number of motors which easily ascend to over two thousand. That is one of the reasons why it is important to maintain an up-to-date database of equipment. In this particular case, such information was obtained through the consultation of the database and through factory surveying, which allowed to cross the information between the different sources, updating the database which is a valuable task in itself.
Technical information of the most relevant motors such has torque, rotation speed, power, current, existence of VSD and type of transmission.	Once again, this information is often difficult to gather. In this particular case, the criteria used for defining the most relevant motors were to consider motors with a power greater than 30kW. All the motors which fell under that category were surveyed.

4.5 - Artificial Lighting

Lighting systems offer easy ways to improve efficiency due to the constant developments in LED technologies and lighting control systems. Even though the consumption of such systems is often low when compared with other equipment, it should not be overlooked. On the following table, it is summarized the information requirements concerning the artificial lighting systems.

Table 4.5 Artificial Lightning Information

Item	Comment
Factory survey depicting the layout of the lighting systems and its characteristics.	Changes in the factory, due to maintenance or malfunction, often mean that the information available is not up-to-date. This is particularly true for the lighting systems, since the activity of changing a lighting bulb when it stops working is not always documented, because it appears as such an insignificant task. Nonetheless, as those situations occur, after several years the information that is present in the database is very different from the reality. In this case, since the power consumption of such systems is relatively small, it was possible to work with the available information, even though it was not precise, but such situations should be avoided and updated databases of equipment should be kept.

	Frequently it is trivial for the employees of the factory to describe the usage of the lighting systems. If that is the case, this information should be provided has it allows to easily analyze the existence of energy waste. If not, this information should be gathered, taking into account the seasonable behavior that the illumination system might follow.
Number of hours in which the illumination is turned on.	

4.6 - Compressed Air Systems

Compressed air is used in such a transversal way in all industries that it is commonly seen as a fourth utility, after electricity, water and gas. If it were a utility it would be the most expensive of all: compressors offer relatively low efficiency, leakage in pipes is common and misuse on behalf of the staff is typical. As such, it is often an area in which efficiency improvement opportunities exist and thus data about this issue is valuable; Table 4.6 summarizes the most important items to provide and comments on them.

Table 4.6 Compressed air systems Information

Item	Comment
Technical data concerning the compressors installed.	The compressors highly impact the efficiency of the compressed air system, as such precise information, including the characteristic curve, is required. If not available in the database, contacting the firm that provided the compressors might be a good option.
Survey of the compressed air pipes, displaying the heights and the existence of air reservoirs.	This survey is very important in the analysis of the efficiency of the compressed air system, but very frequently, due to changes in the productive process, this layout might not exist or be severely out of date. This was the case for SONAE, but due to the high importance of such data, this layout was created.
Technical data concerning the air treatment system employed.	The condition of the input air impacts the efficiency of the compressor, this way it is interesting to analyze which air treatment systems are employed in the factory. Once again, if technical data cannot be found, contacting the firm that provided the system might be the best option.
Requirements for the output compressed air.	Different industries have different requirements for the compressed air. In this case, SONAE wanted to have 6 bar of pressure through the whole compressed air grid; besides, it is also demanded to have the air dry and particle free.

4.7 - Air-Driven Transportation Systems

Pneumatic systems are widely used in industries, especially for transportation of material. Within these systems, there are several aspects which impact the efficiency, being one of the most relevant the impellers used: wrongly chosen geometries or sizes for the impeller lead to excessive consumption. As such, similar to what occurs with the compressed air systems, data on this matter is valuable; Table 4.7, on the next page, summarizes the information requirements and comments on them.

Table 4.7 Air-Driven Transportation Systems Information

Item	Comment
Technical data concerning the ventilators installed.	There are very different sizes and shapes of ventilators, and the operational characteristics of the device highly impact the efficiency of the air-driven systems. As such, the operational curves of the ventilators are valuable information. Once again, if not available in the archive, contacting the company who supplied the ventilators might be a good option.
Survey of the air transportation pipes, displaying the heights and diameters.	Similar to what was commented concerning the survey of the compressed air pipes (4.6), this layout is important but is frequently out of date or inexistent. This was the case for SONAE, which decided to create the survey due to its importance for the project.
Information about the material transported.	The material to be transported affects the possible solutions for the pneumatic system. For example, the kind of impeller to be installed depends on what is being transported: some impellers might stall or quickly degrade due to the impact of the particles transported. As such, it is important to understand which material is going to be conveyed.
Number of hours in which the air-driven transportation systems are being used.	It is useful information in order to understand the power consumption such systems involve, and it is easily obtained by interviewing experienced staff.

4.8 - De-dusting Systems

De-dusting systems might not be present in all sorts of industry, and in some industries they might even be categorized under the air-driven transportation systems, since they are, in fact, air-driven transportation systems. Nonetheless, for the industry currently under discussion, these systems are of such importance that it is even important to distinguish them from other similar systems. Such importance is due to the particularities these systems present in terms of filtering needs, pipe layout, and efficiency improvement opportunities, which are very different from their similar, but different, air-driven systems. Table 4.8 summarizes the information that should be elicited concerning the de-dusting systems.

Table 4.8 De-dusting Systems Information

Item	Comment
Technical data concerning the ventilators installed.	See Table 4.7
Survey of the de-dusting pipes, displaying the heights and diameters.	See Table 4.7
Information about the material transported.	See Table 4.7
Number of hours in which the de-dusting systems are being used.	See Table 4.7

Item	Comment
Technical data concerning the filters.	The filters present in this system might not be the most adequate, which can lead to pressure or speed drops. As such, understanding their behavior is useful.
Type of maintenance performed on the filters.	The type of maintenance helps to characterize the filter and to understand if it is, in fact, the most adequate choice.

4.9 - Energy Management

Under this last topic - energy management - all remaining relevant information concerning the energy policies applied on the factory should be provided. Depending on the site in which the audit is taking place, this might comprise different items. Nonetheless Table 4.9 provides a short list of items that, if available, should be provided.

Table 4.9 Energy Management Information

Item	Comment
Last energy audit.	If available, the access to the last audit performed is valuable as it provides a unique way to acknowledge the energy consumption.
List of energy management practices.	Often, industries have internal policies concerning the practices to manage energy. If available, those should be provided.
Documentation concerning centralized energy management software.	Often, industries have centralized energy management software that aims at reducing the peak consumption and helps in the overview of the productive process. If that is the case, reports of the software should be provided.
Other	Other data, that the staff feels can help the auditors, should also be provided, even if not requested. This might include spreadsheets with analyzed information concerning the energy consumption, "rules of thumb" applied in the factory due to empirical needs, regulation concerning minimum and maximum energy consumption agreed with the energy provider, amongst other.

4.10 - Additional Best-practices

The information provided directly impacts the quality of the energy audit, since it shapes the approach the audit team will take and, consequently, their findings. It can be seen as one of the most critical points.

Often it involves a very high workload for the factory staff, which might lead to an overload of work and thus a poor information gathering. From my experience, having external staff go into the factory and gathering all this information is the best choice because it assures the existence of additional human means to help, preventing the high workload. Furthermore, having an external employee whose sole task is to help in the audit, leads the information gathering effort in a more efficient way as this employee will always judge the information gathered according to its purpose, while, on the other hand, a factory worker who is not fully aware of the purpose and importance of the information, might not disclose important details.

Concerning the actual work done in SONAE, it was accomplished through archive research, factory surveying, contacting providers, and staff interview. In the end, all the previously mentioned items were given an answered, which proved very valuable to the audit team.

Chapter 5

Defining Metering Points

As stated before, an energy audit requires monitoring the energy consumptions to understand the energy behavior and to define measures to improve it. As Robert Kaplan, the founder of the strategic performance tool BSC, said in regard to quality: "If you can't measure it, you can't manage it". The same applies for the energy consumption. Since it would not be feasible to measure the consumption of every single device, there is a need to define which points are the most relevant to monitor, both in terms of electrical energy and thermal energy. The information gathered in the previous chapter is of great importance for that, although it is not enough: a responsible auditor must observe the facility and interview the staff to understand the particularities of the process and to gain a more detailed perspective of the electrical and thermal circuits.

5.1 - Initial Factory Overview

Such a task begins with a meeting with all the stakeholders, which should include factory staff, who will accompany the auditors clearing their doubts and empathizing relevant productive process particularities; senior management, who will manage the overall development of the project according to the company expectations; and the auditors themselves. During the meeting, a schedule should be provided for this work, alongside with resources that might be required - such as special protection equipment or access to, otherwise, restricted areas. According to the experience of the auditors in this project, a typical schedule for such a task, in a complex industrial facility, is of 4 days, but keeping a full working week available for such task is advised as the density of the task might lead to delays or to the need of further studies. As an example, the schedule followed in this project is present in the Table 5.1.

Table 5.1 Example of a schedule for preliminary fieldwork

Days	Actions
Day 1	• Preliminary meeting, culminating in detailed week planning and resource requirements; (~1 morning) • General view of the productive process and all factory sectors; (~1 afternoon)
Day 2	• Identification of the thermal-fluid network, taking into account the pumping motors used and their characteristics; • Identification of the water pumping systems; • Identification of the de-dusting systems network, noticing the motors used and their characteristics; • Identification of the air-driven systems network, noticing the type of impellers used and how they are controlled;
Day 3	• Identification of the compressed air systems network, noticing how the compressors are controlled; • Identification of the steam systems, including the pipeline trails and how the production is controlled; • Identification of exhaust gas uses and pipelines;
Day 4	• Detailed analysis of the electrical circuits, understanding exactly where each equipment is connected; • Identification of the monitoring points currently installed in the facility and the characteristics of the meters;
Day 5	• Interviews with senior factory staff to answer any doubts that might have arose during the development of the actions; • Final meeting where the auditor team shares with senior management some preliminary ideas about the next steps and agrees on a date for the delivery of the report stating which points to monitor.

During the whole 5 days, there are several aspects of which the auditors should take special notice, such has:

- Temperature losses and possible process re-arrangements, which would allow for a better utilization of the thermal energy;
- Process control software, specially VSD control and possible changes concerning the variables acting as input to the software (for example: using temperature instead of pressure as an input);
- Impellers used in the ventilation systems, and their characteristics;
- The efficiency level of the motors present.

This way, at a very initial stage, it is possible for the auditors to shape a strategy into where the biggest energy saving opportunities might be, and present those initial findings to the senior management, in a way that the whole team can collaborate and manage each other's expectations so that the final outcome is exceptional for both parties.

5.2 - Conclusions of the Initial Factory Overview

After the conclusion of this first contact with the factory, the auditors are able to identify, at a very initial stage, potential measures to implement, and provide a document, to the management where they detail their findings. When comparing this method to the more traditional way of developing audits: defining the points to monitor the consumption and, immediately implement the meters, one can notice that this method offers some advantages and disadvantages. On one hand, it's a more expensive procedure, it is more time consuming and requires more dislocations to the factory. On the other hand, it allows for a more detailed planning and to take into account very small details that could, otherwise, be overlooked. Overall, taking into account that the expectation for this project is to determine the smallest details in terms of energy efficiency, it is easy to realize that this method, even with its drawbacks, is strongly advised.

Such was the option in this project: at the end of this stage, the auditors submitted a report to the company to plan what they wanted to monitor, motivating their choice, assuring the approval of all the stakeholders.

5.3 - Implementation of the Meters

Upon having a detailed plan of actions, agreed by all the parties, the actual fieldwork of implementing the meters becomes relatively simple. Nonetheless, a few issues affect the quality of this task. Listed below are several guidelines for a successful implementation of the meters:

- Follow the project-plan strictly, both concerning the location of the equipment and the equipment itself, documenting any unforeseen changes that might arise;
- Follow the security measures presently applied to the company: often the implementation of the meters carries the danger of electrocution or severe burn.
- Use meters which have an acceptable error margin for what they are measuring;

Once the implementation is concluded, the consumption will be monitored during a period which is representative of the load diagram of the equipment. As an example, concerning the project for SONAE Indústria, that period is of 3 months.

Figure 5.1 Picture taken during the audit, portraying the installation of a power quality analyzer

Chapter 6

Measures for Energy Efficient Operations

This chapter details the most relevant measures for increasing the energy efficiency of the operations. Framing this chapter in the overall process, the general audit process includes the following stages:

- Information gathering
- Fieldwork to assess strategic metering points
- Meter installation
- Data gathering
- Data analysis
- Decision about electrical efficiency measures

As such, taking into account that the industry specific data cannot be published due to the confidentiality agreement with SONAE, this next chapter will focus on the decisions about energy efficiency measures. Even though the real data analysis will not be presented, it is useful for the reader to understand how the data was evaluated. Therefore, follows a short overview concerning the features the auditors paid special attention:

- Discovery of energy losses in the transportation of thermal fluids, such as vapor, hot gas or thermal-oil: such losses might comprise huge efficiency improvement opportunities, since improving the thermal isolation of the pipes appropriately or reducing the path of fluids through valves, are simple and cheap measures with high energy gains.
- Detection of situations of no-load electrical operation which can be reduced through better equipment control: situations of equipment operation when there is no actual load are common in facilities with a continuous work flow, such as SONAE, since it is easier to maintain a device continuously operating, than to devise a strategy to turn it off when appropriate. Nonetheless, in equipment with high power consumption, this situation leads to very high energy waste, and increased maintenance needs, and is an aspect which can be easily overcome with suitable control means.
- Finding over/under dimensioned equipment, which leads to poor efficiency: the motors that are currently employed in the factory usual work more efficiently when in a range of 75% to 100% load [10], but often the motors are working bellow that range

to anticipate higher load demands due to productive changes and to reduce maintenance requirements (over-dimensioned motors require less maintenance and have a higher life span). Nonetheless, once again, this faulty choice of the equipment leads to huge energy losses, which can be easily compensated with the insertion of variable speed drives or motor substitution.

- Discovery of non-useful operation of equipment: often, in such large facilities, faulty utilization of factory equipment occurs, since the staff is not always aware of the impact of their behavior. Situations such as un-justified use of compressed air, excessive levels of illumination or continuous use of air conditioner equipment is something easy to change which impacts the consumption.

Keeping the above mentioned factors in mind, and using other criteria that are more specific for the MDF industry, it was possible to formulate several potential measures to implement. These measures were weighted with a cost-benefit analysis, which had mainly two concerns:

- The investment would effectively be pay-off within a maximum period of 10 years, and ideally within a period of 5 or less years.
- The measure would not cause any, or would cause little, disruption on the productive process of the factory.

It is important to notice that in such an analysis, not all benefits are easy to assess due to their nature. For example, it is not easy to access the value of having less thermal losses, since the thermal energy occurs as a "side-effect" of the cogeneration plant in the factory: through the incineration of biomass in the cogeneration plant, thermal energy occurs as a "side-effect" of the production of electricity; as such, since the policy is to always produce the maximum amount of electrical energy, thermal energy is readily available, and reducing its losses might be, or not be, relevant, depending on in which sector it occurs. And when it is relevant, its value is difficult to objectively assess its value. The same happens to the costs, for example, it is not easy to access the value of having the production stopped for several hours to change equipment. As such, part of what is difficult about such a project is making decisions with imperfect and incomplete information. To overcome this issue and to develop a successful project, it is greatly significant that, when deciding upon the efficiency improvement measures, all the stakeholders of the project - auditors, factory staff and senior management - meet to discuss each particular measure in opposition to having it decided only by the senior management, which have a better global view of the installation but lack the detailed operations sensibility that the factories staff can offer.

At the time of delivery of this book, the measures that will be subsequently presented were still being discussed by SONAE's management, and it was still not clear which ones would be implemented. Nonetheless the measures presented in the next sections are considered by all the stakeholders as the most feasible, and will be now motivated under a theoretical point-of-view with enough information for the reader to understand why these procedures are relevant.

6.1 - Preventive Maintenance Plan for the Vapor Pipeline

Due to the large extension of the vapor network, and due to the high amount of devices that use the vapor of low and medium pressure, there are several vapor leaks that, as a whole, account for an extremely high energy loss. The preventive maintenance is an important tool to prevent vapor losses and, consequently, energy losses. In particular, the steam traps should be checked often through the analysis of its performance.

Furthermore, when spontaneous vapor leakage is found, it should be addressed with corrective measures as soon as possible, in order to allow the normal productive process (without having the need to resource to natural gas to provide the lacking thermal energy) and avoiding energy losses. In this facility, several vapor leaks were found. Some of the biggest leaks were found in the refiner, in the air grader and in the dryer of the MDF production line 2. Furthermore, the steam trap of the MDF production line 1 is severely degraded and, as such, it presented very large losses.

As a result of the metering process, it was possible to quantify the losses as being:

- Dryer of MDF, line 1: 3100MWht/year
- Leak in the boiler 1: 140 MWht/year

Taking into account the mentioned values, the calculated potential for energy savings is as high as 6.200 MWht/year. The investment for application of this measured depends on which repairs are chosen to be performed, nonetheless it is estimated an investment of 200.000€ per year, which includes the substitution of the steam trap.

6.2 - Optimize the Operation of the Wood Chipper

The wood chipper is a device which has very high power consumption and a discrete operation behavior. It is a device that, due to its high consumption, is already closely followed and measures, such as restraining its operation to hours in which the electricity is less expensive, are already implemented.

Nonetheless, from the measurements obtained it is possible to see that, in certain periods, the wood chipper is working with no load, meaning that it is not performing any action because there is no wood to be turned into chips, but it is still consuming a considerable amount of electricity.

As such, it is advised that when it is possible to foresee long stops (superior to 1 hour) the equipment is shut down. The predicted energy savings with this measure are of about 11MWh/year, with no significant cost associated.

6.3 - Optimize the Operation of the Refiner

The refiner possesses the motor with the highest nominal power at around 4MW. It accounts for a large part of the electrical consumption of the whole factory, and presents a continuous operating behavior, except when the productive processed is stopped. These stops are mainly due to unexpected malfunctions in the production line that lead to unscheduled stops. Analyzing the load diagrams gathered with the meters, one can notice that if the refiner is stopped when there are issues with the production line it would be possible to save around 130kWh/h and 150kWh/h, for the refiner 1 and 2 respectively.

Through the Application of this measure it would be possible to reduce the consumption of energy around 65MWh/year, with no significant cost.

6.4 - Insertion of a Damper in the De-dusting System Pipeline

The main de-dusting system uses a ventilator with a 55kW AC motor for each MDF production line. Concerning the wood- veneer surfacing, there are autonomous de-dusting systems for both line 1 and 2. The wood-veneer surfacing line has several stops during the day since some parameters have to be adjusted according the product that is being created. Therefore, the possibility of installing a damper in the main pipeline was studied, in order to automatically regulate the air flow in the conduct, according to the operation of the sanding machines.

The damper is able to reduce the size of the conduct, which reduces the air flow and, consequently, the energy needed for the ventilation.

The application of this measure in both 55kW motors would reduce energy consumption in about 16 MWh/year, and would comprise a cost of around 4.000€

6.5 - Insertion of a VSD in the thermal fluid pump of the wood veneer surfacing press

The wood veneer surfacing process does not usually work on weekends and, therefore, it would be profitable if it were possible to reduce the flow of thermal fluid during these times. Furthermore, the distance between the consumption of this thermal energy and its production is one of highest possible, since the production site and the consumption site are in opposite parts of the factory the pipes that connect the two sites are around 200 meters.

For the purpose of circulating the thermal fluid there is a pump with a nominal power of 30kW. This pump works continuously even when there is no use for the thermal fluid since the presses are not functioning. Since it would not be convenient to turn off the pump when stops in the presses occur, it is advised the insertion of a VSD in the pump in order to automatically adjust the flow of thermal fluid with the needs of the wood veneer surfacing presses.

The application of this measure is estimated to save 61 MWh/year and the cost for the installation is estimated to be circa 3.000€.

6.6 - Reduction of the Compressed Air Leakage

As it was previously stated, the compressed air production encompasses very high costs in electrical spending and, as such, the detection and repair of compressed air leakage should constitute a priority of the maintenance team [11].

The maintenance team should devise a plan to insert control valves in the compressed air pipes, allowing that in periods in which certain equipment is not working - such has when the wood-veneer surfacing lines are shut down - the compressed air only goes through the least amount of pipes. Furthermore, awareness campaigns for the importance of reducing compressed air use should be put in practice focusing on key operators of the factory.

Additionally, since the air leakages might be difficult to find in the working environment of the factory, which is characterized by high levels of noise and the existence of particles that partly impair the vision, the purchase of the an ultrasonic flow meter, which allows to see the leakage of compressed air, is advised.

Taking into account the results from the metering stage, it is estimated that the reduction of the air leakage would save 650MWh/year, with a cost of around 3.000 € for the purchase of the ultrasonic flow meter.

6.7 - Improvement of the Ventilation of the Compressors Room

The room where the main compressors are found does not possess forced ventilation, having only grids that allow the circulation of the air. The temperature in this room, especially in the summer, is very high due to the insufficient means of air renovation. Taking into account that the air admitted by the compressor should be as cold as possible, this situation constitutes an efficiency handicap.

This way, this situation should be changed with the insertion of a ventilator, which could be helical, that would be big enough to allow for a proper air renovation. According to the technical data sheet of the compressors, a decrease of 4°C in the admission air will result in a 1% efficiency improvement.

Taking into account that, under normal operation, only one compressor is working, it is predicted an energy saving of 13MWh/year, already taking into account the extra energy consumption a ventilator would impose. The investment cost is about 1.000€.

6.8 - Insertion of a Variable Voltage System for the Artificial Lighting

The artificial illumination of the industrial complex is mainly done resourcing to mercury vapor lighting bulbs (83%). These kind of lighting bulbs, after being turned on, require a period of time to provide the nominal level of illumination. During this time they have to be under the nominal voltage of the network, at 230V. Nonetheless, after the bulbs reach the maximum illumination stage, they do not need to be under the tension of 230V. Currently, there are systems that allow the reduction of the tension down to a limit specified by the producer of the lightning solution. The main advantages that this solution are:

- Reduction of the power consumption between 20% and 40%.
- Longer life-time of the lighting bulbs, reducing maintenance costs;
- Stable voltage in the illumination circuits.

Naturally, this solution would impose a reduction of the illumination capabilities of each bulb, but taking into account the amount of bulbs in the complex and the good level of illumination currently imposed, the difference would be barely noticeable.

Taking into account the number of luminaires, the power of each lighting bulb, and the normal working behavior of the illumination systems, it is calculated that the installation of this system would allow a reduction in the consumption of electricity of around 13MWh/year. The total investment for the implementation of the system is around 3.000€.

6.9 - Implement Gas Burner Automatic Control

Upon noticing high natural gas consumptions, the auditors focused on pin-pointing the causes of such consumption. It was noticed that, concerning the fiber drying stage - when the exceeding water content of the fiber is thermally extracted with hot air- that when thermal energy is not available from the cogeneration plant, natural gas is used instead. Even though there is an automatic control for the inlet valve that allows the admission of the gas, this control is done manually, because the automatic control is not properly calibrated, and to overcome the problem the operators decided to control the admission of gas according to their own expertise of the process.

Upon noticing the situation and properly inspecting the control software, the factory staff calibrated the software correctly which resulted in a 25% reduction on the consumption of natural gas on the first month of usage, which is translated in a reduction of 5500 MWh/year, without any investment costs!

6.10 - Rejected Measures for efficiency increase

In the process of pin-pointing feasible opportunities to increase the efficiency, and since the more usual measures (such as changing the impellers of the ventilators, inserting VSD into motors or substituting less efficient equipment) had already been implemented, several measures which were more challenging were suggested. In this sub-chapter some of the measures that were discarded will be detailed, explaining both why they were suggested and why they were not accepted.

6.10.1 - Installation of a photovoltaic panel

This measure advised the installation photovoltaic panel of 150kWp, with an area of 2.500 m2 and an estimated production of electricity of 276MWh per year. This estimation was based on actual measures of solar exposition as well as different parameters inherent to photovoltaic feasibility studies. The investment for such a project was estimated to being around 880.000€.

This measure was declined by senior management mainly due to two reasons:
- The pay-off time would be higher than 10 years;

- The current economic reforms on energy polices could change the profitability of the installation if the prices paid by the electricity service provider were to decrease in the future.

As such, even though this idea is feasible and would lead, in the long-term, to savings, it was ultimately declined.

6.10.2 - Renewing the Thermal Isolation of the Thermal-fluid pipes in the Glue Tank

This measure advised the renovation of the thermal isolation of the thermal fluid pipes to reduce losses. Taking into account the thermal-fluid normal temperatures (between 75°C and 80°C), it was estimated that renewing the isolation would result in savings of 308MWh/year, with an estimated cost of 3.000€. At a first look, it seems like a very profitable measure but since the thermal energy, in this particular section, is a result of the incineration of biomass, and since that incineration will be producing the same amount of thermal energy regardless of its use in the system, there is, in fact, no advantage in improving the isolation and the thermal losses.

Ultimately this study will be kept as a feasible measure to be taken if at any point there is a lack of thermal energy.

6.11 - Overview of the Measures

Overall the measures elicited are very interesting. They build on the previous efficiency improvement work done in the factory, and if replicated in the remaining 25 factories owned by the group, might result in truly important savings for the company. In the next chapter a financial analysis of the measures will be presented, using energy prices that are realistic but not exactly what is paid by SONAE as it pays a special price due to its' private agreements with the energy providers, which cannot be disclosed.

Chapter 7

Expected Outcome of the Implementation

This chapter presents the financial analysis of the previously advised measures. It is important to remember that confidentiality agreements do not allow to publish the actual energy prices paid by SONAE; so the following reference values will be used: 60€ per electrical MWh; 30€ per natural gas MWh, and 20€ per MWh of thermal energy, which are realistic values for the energy prices. Also concerning the specific consumption of the operations, the real values cannot be published but, in order to provide a solid idea of the accomplishments to the reader, the consumptions will be provided after being multiplied by a numerical factor.

Table 7.1 Energy prices

	Source of Energy	Price €/MWh
1.	Electrical energy	60.0
2.	Natural Gas	30.0
3.	Thermal Energy	20

7.1 - Summary of Energy Efficiency Measures

To frame this next chapter into the previously accomplished work, table 7.1 summarizes the energy efficiency measures that were seen as the most feasible, recalling the previously presented figures concerning the costs and potential savings, and calculating the pay-back time and implementation year. It is important to notice that, when in case of uncertainty concerning the cost of the measure or the potential savings, the criteria used was to assume the most pessimistic scenario, in order to protect the interests of SONAE.

Regarding the implementation year, this is a choice that, at the time of delivery of this dissertation, was still not ultimately decided. Nonetheless, the criterion used to decide upon when the measures were to be implemented was based on the pay-back time (the shortest the period for the pay-back the more urgent the measure became) and, the eventual disruption of the productive process that the measure could impose (e.g. measure 4. Requires that the de-dusting system is momentarily not working, as such it is better to implement this change in a period in which the de-dusting system has a scheduled maintenance).

Table 7.2 Summary of energy efficiency measures

	Measures	Savings		Investment	Pay-back	Implement.
		MWh/year	€/year	€	years	year
1.	Preventive maintenance plan for the vapor pipelines	6.200	124.000	200.000	1,62	2012
2.	Optimize the operation of the wood chipper	11	660	-	0	2012
3.	Optimize the operation of the refiner	65	3900	-	0	2012
4.	Insertion of a damper in the de-dusting system pipeline	16	960	4.000	4,17	2013
5.	Insertion of a VSD in the thermal fluid pump of the wood veneer surfacing press	61	3660	3.000	0,82	2013
6.	Reduction of the compressed air leakage	650	39000	3.000	0,08	2012
7.	Improve the ventilation of the compressors room	13	780	1.000	1,29	2013
8.	Insertion of a variable tension system for the lightning systems	13	780	3.000	3,85	2013
9.	Implement Gas Burner Automatic Control	5500	165000	-	0	2012
	Overview of all the measures	**12.529**	**338.740**	**214.000**	**0,631753**	

As the reader can notice, if all the measures are implemented, even with pessimistic criteria a pay-back time of around 0,63 years, meaning 8 months, is expected, and savings of around 338.000€ are expected for the following years. If these measures were able to be replicated in the remaining 25 factories of SONAE, it would be possible for the all group to save around 8.000.000€ per year, but naturally, further studies would have to be conducted to understand which measures could be replicated and new financial analysis would have to be performed to take into account the energy prices in the geographic area in which the remaining factories operate.

7.2 - Characterization of the Current Energy Consumption

In this sub-chapter, a characterization of the current energy consumptions will be presented, so that the reader can compare the actual situation, with the expected situation after the implementation of the energy efficiency measures (presented under the 7.3). Once again, it is important to notice that, due to the confidentiality agreement; the figures presented are not completely accurate, having been multiplied by a factor. Nonetheless, they portray a realistic overview of the consumption.

7.2.1 - Electrical Energy

The average consumption of electrical energy is around 7MWh per month, without taking into account months in which there is maintenance. It is the most widespread energy in the facility and the biggest cost driver. Notice further details on Table 7.3.

Table 7.3 Electrical Energy Consumption in 2011

Electrical Energy		
Period	Quantity	
	MWh	€
January 2011	7.025	421.509
February 2011	7.106	426.372
March 2011	7.701	462.050
April 2011	7.070	424.184
May 2011	7.545	452.689
June 2011	7.187	431.234
July 2011	7.341	440.473
August 2011	3.369	202.154
September 2011	7.332	439.926
October 2011	7.648	458.889
November 2011	6.279	376.714
December 2011	4.986	299.159
TOTAL	80.589	4.835.353

7.2.2 - Thermal Energy

Thermal energy is very important for the productive process, particularly in cooking the wood chips, drying the wood fiber and heating the presses. Notice on Table 7.4 the large thermal energy consumption that such a facility presents.

Table 7.4 Thermal Energy Consumption in 2011

Period	Quantity	
Thermal Energy		
	MWht	€
January 2011	6.419	128.372
February 2011	4.721	94.419
March 2011	4.302	86.047
April 2011	4.174	83.488
May 2011	3.337	66.744
June 2011	3.384	67.674
July 2011	4.767	95.349
August 2011	2.047	40.930
September 2011	4.465	89.302
October 2011	3.523	70.465
November 2011	3.674	73.488
December 2011	2.512	50.233
TOTAL	**47.326**	**946.512**

7.2.3 - Natural Gas

Energy whose source is natural gas is the least used in the process, but still an important cost driver. Furthermore, recent increases in prices make the need to consume this resource rationally more and more important.

Table 7.5 Natural Gas Consumption in 2011

Period	Quantity		
Natural Gas			
	m³	MWh	€
January 2011	149.184	1.800	54.000
February 2011	143.253	1.657	49.710
March 2011	178.377	2.154	64.620
April 2011	168.379	2.013	60.390

May 2011	146.023	1.699	50.970
June 2011	96.788	1.166	34.980
July 2011	110.414	1.244	37.320
August 2011	71.117	827	24.810
September 2011	132.288	1.482	44.460
October 2011	133.837	1.586	47.580
November 2011	193.125	2.181	65.430
December 2011	197.853	2.161	64.830
TOTAL	**1.720.638**	**19.970**	**599.100**

7.2.4 - Specific Consumption of the MDF product

The specific consumption is metric that relates the production of the MDF with the use of electrical energy. It is a metric used to perform analysis between the different factories which produce the same product but in different amounts, and is helpful to highlight the energy consumption variations. Furthermore, it is used as a managing tool to set objectives on a corporate level that factories should follow. Table 7.6 presents the specific consumption in the different months of 2011.

Table 7.6 Specific consumption of the MDF production in 2011

	Specific Consumption		
Period	MDF Production m3	Electrical Energy kWh	Specific Consumption KWh/m^3
January 2011	21.637	7.025.155	324,68
February 2011	22.035	7.106.195	322,50
March 2011	24.648	7.700.826	312,44
April 2011	23.025	7.069.727	307,05
May 2011	24.109	7.544.824	312,95
June 2011	22.861	7.187.235	314,38
July 2011	23.298	7.341.211	315,09
August 2011	9.407	3.369.238	358,18
September 2011	24.068	7.332.094	304,65
October 2011	24.462	7.648.150	312,65
November 2011	18.031	6.278.574	348,20
December 2011	13.421	4.985.986	371,51
TOTAL	**251.002**	**80.589.215**	**321,07**

7.2.5 - Overview of all the energy consumption

To summarize the energy consumption of the facility, the energy provided by the different sources is converted to the unit *toe*, using the conversion factors documented in the Portuguese legislation, and presented in Table 7.7 Conversion Factors [4]. The result is presented in Table 7.8 Overview of all the energy consumption and allows the comparison of consumption between the different energy sources.

Table 7.7 Conversion Factors

Conversion Factor		
Electrical Energy	0,215	toe/MWh
Thermal Energy	0,086	toe/MWht
Natural Gas	0,00082	toe/m3

Table 7.8 Overview of all the energy consumption

Energy Overview				
Period	Electrical Energy	Thermal Energy	Natural Gas	Total
	toe	toe	toe	toe
January 2011	1.510	552	122	2.185
February 2011	1.528	406	117	2.051
March 2011	1.656	370	146	2.172
April 2011	1.520	359	138	2.017
May 2011	1.622	287	120	2.029
June 2011	1.545	291	79	1.916
July 2011	1.578	410	91	2.079
August 2011	724	176	58	959
September 2011	1.576	384	108	2.069
October 2011	1.644	303	110	2.057
November 2011	1.350	316	158	1.824
December 2011	1.072	216	162	1.450
Total	17.327	4.070	1.411	22.808

Clearly, and as stated previously, electrical energy is the most used form of energy followed by thermal energy and natural gas.

7.3 - Predicted Consumption after the Implementation of the Measures

Having characterized the present consumptions of the factory, this chapter presents the future consumptions after the implementation of the advised measures, in order to compare both situations. Besides assuming that the previously savings estimates are correct, this prediction possesses two assumptions that are important to notice:

- The energy saving each measure provides is the same in each month;
- The energy prices remain constant;

Both assumptions are realistic and should portray a precise prediction.

7.3.1 - Electrical Energy

Concerning the electrical energy, measures from 2 to8 in Table 7.2 help reduce the consumption in 830MWh/year, which is reflected in a decrease of around 70MWh per month. Using the previously presented figures for the electrical consumption, and taking into account this decrease, the consumption will be as shown in table 7.9.

Table 7.9 Electrical Energy Consumption After the implementation of the advised measures

Period	Electrical Energy	
	Quantity	
	MWh	€
January	6.955	417.309
February	7.036	422.172
March	7.631	457.850
April	7.000	419.984
May	7.475	448.489
June	7.117	427.034
July	7.271	436.273
August	3.299	197.954
September	7.262	435.726
October	7.578	454.689
November	6.209	372.514
December	4.916	294.959
TOTAL	79.749	4.784.953

Comparing this situation with the present, one can notice a decrease in costs of around 50.000€ per year. But it should also be taken into account that the measures advised also help reduce maintenance needs of equipment and, as such, savings could be even greater.

7.3.2 - Thermal Energy

Concerning the thermal energy, measure number 1 helps reduce the consumption in 6200 MWh_t per year, which is around 515 MWh_t per month. Once again, taking into account the previously presented figures, it is possible to estimate consumptions, as shown in table 7.10.

Table 7.10 Thermal Energy Consumption After the implementation of the advised measures

Thermal Energy		
Period	Quantity	
	MWht	€
January	5.903	118.052
February	4.205	84.099
March	3.786	75.727
April	3.658	73.168
May	2.821	56.424
June	2.868	57.354
July	4.251	85.029
August	1.531	30.610
September	3.949	78.982
October	3.007	60.145
November	3.158	63.168
December	1.996	39.913
TOTAL	41.134	822.672

Comparing this situation with the present, a decrease in costs of around 124.000€ per year can be noticed. Furthermore, taking into account the steady increase in energy prices, this saving could be even more relevant in the future.

7.3.3 - Natural Gas

Regarding the natural gas, it is impressive to notice that a simple control software calibration, with no investment cost other than a few hours of staffs' time, could lead to savings of around 5500 MWh per year, meaning around 458 MWh per month. Using the same method as before, the consumption can be estimated as shown in table 7.11.

Natural Gas		
Period	Quantity	
	MWh	€
January	1.341	40.230
February	1.198	35.940
March	1.695	50.850

April	1.554	46.620
May	1.240	37.200
June	707	21.210
July	785	23.550
August	368	11.040
September	1.023	30.690
October	1.127	33.810
November	1.722	51.660
December	1.702	51.060
TOTAL	14.462	433.860

Comparing this situation with the present, one can notice a decrease in costs of around 165.000€ per year.

7.3.4 - Overview of the differences

A global perspective of the expected differences is now presented, through summarized charts in Figure 7.1 and Figure 7.3.

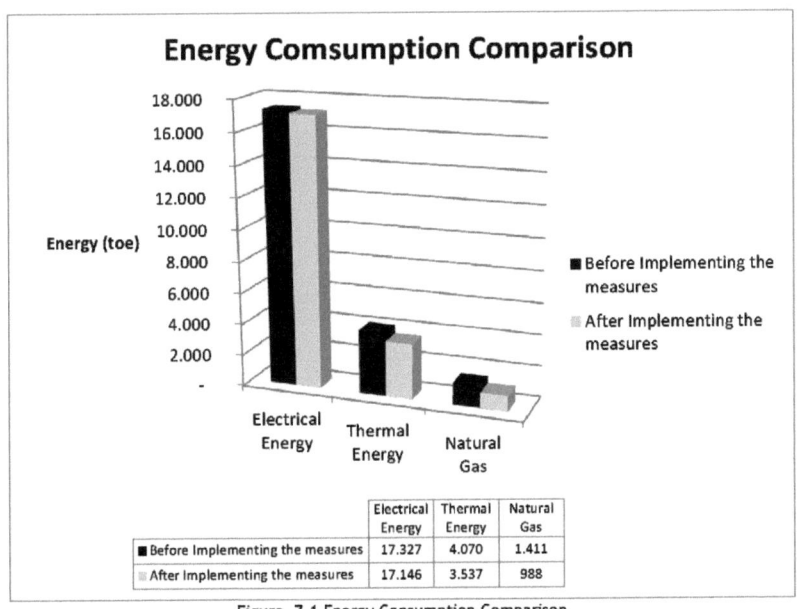

	Electrical Energy	Thermal Energy	Natural Gas
Before Implementing the measures	17.327	4.070	1.411
After Implementing the measures	17.146	3.537	988

Figure 7.1 Energy Consumption Comparison

Clearly the biggest shift in consumption is seen in the natural gas, where a decrease of around 30% is expected, followed by thermal energy with a decrease of around 14%, and followed by electrical energy with a small decrease of around 2%.

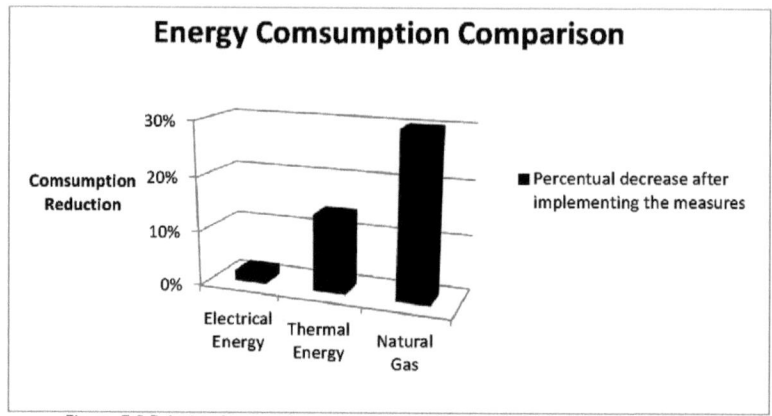

Figure 7.2 Relative decrease in energy consumption after implementing the measures

Concerning the energy costs, the same relative decrease is experienced, leading to the savings summarized on Figure 7.3 Energy Spending Compariso

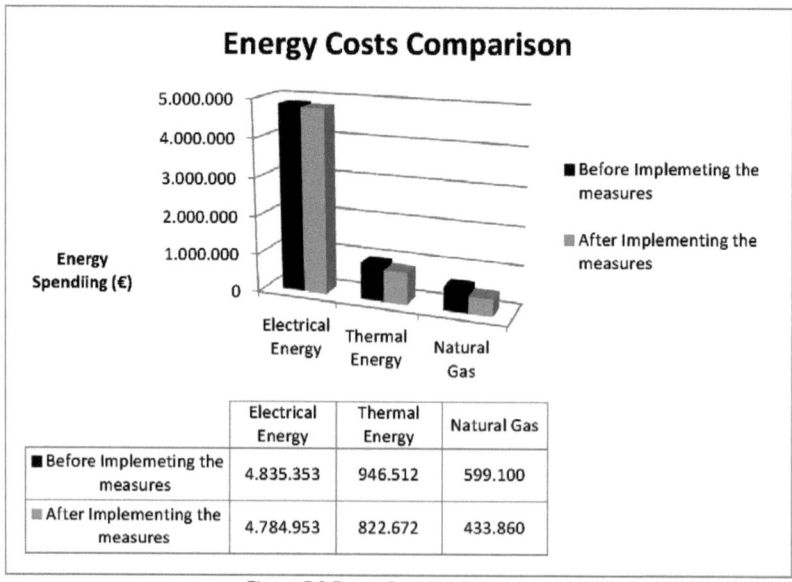

	Electrical Energy	Thermal Energy	Natural Gas
■ Before Implemeting the measures	4.835.353	946.512	599.100
▨ After Implementing the measures	4.784.953	822.672	433.860

Figure 7.3 Energy Spending Comparison

From a purely financial stance, an impressive business opportunity is noticeable when calculating the internal return rate, which amounts to 148%. Even when considering a cost of capital of 10%, a modified internal return rate calculation provides a rate of 74%. This way, from a purely financial point-of-view, it is a very interesting investment.

Table 7.11 Financial Analysis of the advised measures

Financial Analysis					
Year		0	1	2	3
Inflation	2%		1	1,02	1,0404
Savings		0	338.740,00	338.740,00	338.740,00
Costs		214.000,00	0	0	0
Cash-Flow		-214.000 €	338.740 €	338.740 €	338.740 €
Cost of Capital	10%				
Weighted Cash-Flow		-214.000 €	307.945 €	279.950 €	254.500 €
NPV	628.396 €				
IRR	148%				
MIRR	74%				

And in a broader overview, these measures also contribute to more efficient productive process, which is translated in a lower cost of production and a bigger profit margin in each product, which allows the company to maintain a leading position within the wood-products industry either by lowering its product prices or by having higher profits, which are reinvested in the expansion of the business.

Chapter 8

Conclusions and Future Work

This final chapter summarizes the whole project, highlights its main advantages and disadvantages and refers future works that can be undertaken to keep pursuing a more efficient production process.

8.1 - Overview of the Problem and final analysis of results

The main expectation of this project was to build on the previous energy efficiency work that had been accomplished and further reduce consumptions and costs, as such it can be seen as very successful: it accomplished to discover measures which reduce the consumption in around 1100 toe per year, which translates in around 330.000€ per year, with a payback of 8 months. From this perspective, no doubts arise, it is a highly valuable project, especially if it can be replicated in the remaining factories.

On the other hand, there was also the expectation of going into the smallest energy details. Energy audits and efficiency improvement work had been done in the past, and it was believed that the factory was, at the moment, already very efficient. As an example, SONAE was selected as a "Most Efficient Company" at the Energy Efficiency Awards Portugal 2010 [12]. This expectation was, from my perspective, not completely met. It was possible to find measures that did not require particularly strong technical or creative skills, nor did they require cutting-edge new practices. Nonetheless, the measures still have a very significant impact on consumptions. As such, it is possible to conclude that if SONAE is a "Most Efficient Company" and still has improvement opportunities present, other companies must have a huge potential for improvement, and can reduce their consumptions and costs greatly.

Concerning the development of the project itself, it was very positive. The teamwork, in a team which had people of very different companies, was impressive, and in any particular work day a high amount of work was achieved. During this process, while gathering information for the audit, it was possible to reorganize the archives, to update information concerning the equipment, and to update existing layouts of different pipelines.

The main difficulties found had to do with lack of organization, both in terms of documents and in terms of schematics of the different circuits. This problem can be understood when working in such a big facility, which suffered improvements not always documented in the best way, and was overcome due to the efforts of the team.

Overall, from my perspective, it is a project that should be replicated in industries throughout the world, as these projects undoubtedly create value for any company and help meet political goals concerning pollution and energy consumptions, improving society in its whole.

8.2 - Future works

Future projects with the same goal, improving energy efficiency, should be performed as there are still opportunities to be found. From my perspective, these projects would now have to enter through new approaches, aiming at finding changes in the productive process to better utilize energy. Furthermore, less profitable measures should also be considerable if they prove to be feasible on the long-term and, for that analysis, not only the reduction of energy consumption needs to be accounted for, but also the reduction of maintenance requirements.

Nonetheless, it is easy to understand that such an approach would require large funding and that the tough economic situation Europe is experiencing does not allow for such a spending at the moment. As such, I feel that even though future projects are in order, SONAE is working in the correct direction concerning this issue and should be seen as an example to be followed.

References

[1] "Action Plan For energy Efficiency," [Online]. Available: http://europa.eu/legislation_summaries/energy/energy_efficiency/l27064_en.htm. [Accessed 10 Junho 2012].

[2] F. Sánchez, "Manual de boas práticas de eficiência energética.," [Online]. Available: http://www.bcsdportugal.org/files/496.pdf. [Accessed 19 Junho 2012].

[3] "Sonae Indústria Company Overview," [Online]. Available: http://www.sonaeindustria.com/page.php?ctx=1,0,17. [Accessed 10 Junho 2012].

[4] *Decreto-Lei nº 71/2008, de 15 de Abril, do Sistema de Gestão de Consumos Intensivos de Energia,* 2008.

[5] "A Universidade do Porto em Números," Reitoria da Universidade do Porto, 2009. [Online]. Available: http://sigarra.up.pt/up/web_base.gera_pagina?p_pagina=122350. [Accessed 19 Junho 2012].

[6] *Sonae Indústria PCDM Mangualde Brochure,* 2012.

[7] C. Gaspar, Eficiência Energética na Indústria, Vila Nova de Gaia, 2004.

[8] T. G. o. t. H. K. S. A. Region, Guidelines on Energy Audit, Hong Kong, 2007.

[9] D. Wulfinghoff, Energy Efficiency Manual, Energy Institute Press, 2000.

[10] U. D. o. Energy, Improving Motor and Drive System Performance, 2008.

[11] U. D. o. Energy, Improving Compressed Air System Performance, 2008.

[12] "Sonae Industria is awarded with an Energy Efficiency Award," [Online]. Available: http://www.sonaeindustria.com/file_bank/press/announcements/PR20110427-Eng.pdf. [Accessed 13 Junho 2012].

[13] "Relatório de Progresso Anual 2010," Mangualde, 2011.

[14] F. Leitão, "Energy Efficiency Program - 8th Forum Meeting," SONAE Indústria, 2012.

Printed by Books on Demand GmbH, Norderstedt / Germany